CORPUS
PONDERUM ANTIQUORUM ET ISLAMICORUM

Turkey 1

CORPUS
PONDERUM ANTIQUORUM ET ISLAMICORUM

Turkey 1

THE COLLECTION OF KLİMA PLUS IN SİLİFKE MUSEUM

Greek, Roman, Byzantine and Islamic Weights

by

Oğuz TEKİN – Ali M. MERZECİ

TURKISH INSTITUTE OF ARCHAEOLOGY

ISTANBUL
2012

Corpus Ponderum Antiquorum et Islamicorum is a project of Turkish Institute of Archaeology.

Abbreviated Form of the Project Name: *CPAI*

Director of Project: Oğuz Tekin

To the users:
Corpus Ponderum Antiquorum et Islamicorum aims only to present a rich source of material relating Greek, Roman, Byzantine and Islamic weights. Dates are approximate and should be used cautiously. Catalogue entries contain only short descriptions and they require further studies.

Corpus Ponderum Antiquorum et Islamicorum. Turkey 1. The Collection of Klima Plus in Silifke Museum. Greek, Roman, Byzantine and Islamic Weights, 46 plates.
ISBN: 978-605-4701-00-1

Catalogue and photographs by Oğuz Tekin

Frontispiece: Cat. no. 8

Published by Turkish Institute of Archaeology
Istiklâl Cad. Merkez Han, No: 181 Kat: 2
34435 Beyoğlu - Istanbul, Turkey
Tel: 0090-212-292 09 63 – 514 03 97
www.turkinst.org

Typesetting and page setup: Zero Prod. Ltd., Istanbul, Turkey

Printing and binding: MAS Matbaacılık A.Ş.
Hamidiye Mah. Soğuksu Cad. No. 3 Kağıthane - Istanbul, Turkey
Tel: +90 (212) 294 10 00 Fax: +90 (212) 294 90 80 info@masmat.com.tr
Certificate no: 12055

EU distributor: Zero Prod. Ltd, Abdullah Sokak 17, Taksim 34433, Istanbul, Turkey
Tel: +90 (212) 244 75 21 Fax: +90 (212) 244 32 09
E.mail: info@zerobooksonline.com Web site: www.zerobooksonline.com

CONTENTS

To the memory of Prof. Dr. Ali M. Dinçol

PREFACE

Corpus Ponderum Antiquorum et Islamicorum (=CPAI) is a project initiated in 2008 by the Turkish Institute of Archaeology. Starting point of the project is the rarity of scientific publications about the ancient and Islamic weights when it is compared with other archaeological material and the necessity that the theme of weights should be moved to the scientific platform. In the scope of the project, the commercial weights in the museums of Turkey, which comprises the geography of Anatolia and Turkish Thrace, were studied and formed a digital archive. Since the aim of the project is to present a rich source of material relating to the Greek, Roman, Byzantine and Islamic weights, the weights in the museums and private collections in Turkey will be published as separate volumes of CPAI as well as articles in the following years.

The weights in this catalogue were collected through acquisition from Mersin and its vicinity (in ancient Cilicia) by Ali M. Merzeci, a conscious person to cultural heritage, and financed by Klima Plus Company in Mersin, and were finally donated to the Silifke Museum where ancient Seleucia ad Calycadnum lies. The catalogue was prepared by me.

I would like to express my thanks to Ali M. Merzeci and to Klima Plus Company to whom we owe this modest collection in Silifke Museum. I would like to thank İlhame Öztürk, the directress of the Silifke Museum, who facilitated my works in the museum and Arif Yacı, one of the MA students of me who helped me during my work in the museum. My special thanks also go to John H. Kroll and Bendeguz Tobias, for their kind suggestions and comments; to Chris Entwistle, for sending me his draft catalogue of Byzantine weights in the British Museum; to İ. Günay Paksoy and Muhammet Yücel for their comments on arabic inscriptions on some weights (cat. nos. 95, 104, 107 and 115); of course the responsibility for the mistakes belongs to me. I would also like to thank the Managing Board of the Turkish Institute of Archaeology in the person of its President late Prof. Dr. Ali M. Dinçol, who gave enormous support to the project of CPAI but sadly we lost him very recently. Further thanks are due to Ahmet Boratav of Ege Publications for his efforts in preparing the catalogue for publication in such a short time.

Oğuz Tekin
tekinotekin@yahoo.com
www.academia.edu

ABBREVIATIONS

Balog 1970
P. Balog, "Islamic Bronze Weights from Egypt", *Journal of the Economic and Social History of the Orient* XIII (1970), 233-256.

Bendall 1996
S. Bendall, *Byzantine Weights. An Introduction*, London, 1996.

Binsfeld 1990
W. Binsfeld, "Römische Gewichte in Trier", *Trierer Zeitschrift* 53 (1990), 281-290.

Boucher 1971
S. Boucher, *Vienne. Bronze antiques*, Paris, 1971.

Byzanz. Licht aus dem Osten 2002
Byzanz. Das Licht aus dem Osten. Kult und Alltag im Byzantinischen Reich vom 4.bis 15. Jahrhundert. Katalog der Ausstellung im Erzbischöflichen Diözesanmuseum Paderborn 2001 (ed. C. Stiegemann), Mainz, 2002.

Corti – Pallante – Tarpini 2001
C. Corti – P. Pallante – R. Tarpini, "Bilance, Stadere, Pesi e Contrappesi nel Modenese", *Pondera, Pesi e Misure nell'Antichita* (ed. C. Corti and N. Giordani), Museo della Bilancia, 2001.

Crowfoot – Crowfoot – Kenyon 1957
J.W. Crowfoot – G.M. Crowfoot – K.M. Kenyon, *The Objects from Samaria*, London, 1957.

Custurea 1995-1996
G. Custurea, "Cateva ponduri mediavale descoperite in Dobrogea", *Pontica* XXVIII-XXIX (1995-1996), 323-328.

Custurea 1998
G. Custurea, "Noi ponduri mediavale din Dobrogea", *Pontica* XXXI (1998), 307-308.

Custurea 2002-2003
G. Custurea, "Ponduri antice și mediavale descoperite in Dobrogea", *Pontica* XXXV-XXXVI (2002-2003), 439-443.

Davidson 1952
G.R. Davidson, *Corinth. Vol. II. The Minor Objects*, Princeton, New Jersey, 1952.

Decloedt 1914
A. Decloedt, "Note sur des poids grecs et byzantins du Musée Biblique de Sainte-Anne", *Revue Biblique* 11 (1914), 549-55.

Dieudonné 1931
A. Dieudonné, "Poids du bas-empire et byzantins des Collections Schlumberger et Froehner et de l'ancien fonds du Cabinet", *Revue numismatique* 34 (1931), 11-22.

Die Welt von Byzanz 2004
Die Welt von Byzanz-Europas östliches *Erbe. Glanz, Krisen und Fortleben einer tausendjährigen Kultur, 22. Oktober 2004 bis zum 3. April 2005 in der Archäologischen Staatssammlung München - Museum für Vor- und Frühgeschichte* (Ed. L. Wamser), Stuttgart, 2004.

Draganov 1994
D. Draganov, "Bas antiques poids de Cabyle et hinterland", *Izvestija na Muzeite na Jugoiztocna Bulgarija* 17 (1994), 141-148.

Dürr 1964
N. Dürr, *Catalogue des poids byzantines. Musee d'Art et d'Histoire de Geneve*, Geneve, 1964.

Gorny und Mosch
Gorny und Mosch Giessener Münzhandlung.

Hadat 2007
S. Hadat, "Weights from the Early Roman Period at Ramat Hanadiv", *Israel Exploration Journal* 57 (2007), 208-210.

Hendin 2007
D. Hendin, *Ancient Scale Weights and Pre-Coinage Currency of the Near East*, New York, 2007.

Holland 1986
L. Holland, "Islamic Bronze Weights from Caesarea Maritima", *ANSMN* 31 (1986), 171-201.

Holland 2009
L. Holland, *Weights and Weight-like Objects from Caesarea Maritima*, Hadera, 2009.

Isler 1999
H.P.Isler, "Pesi di bronzo islamici dagli scavi di Monte Iato (Sicilia)", *Numismatica e Antichita Classische* XXVII (1998), 345-369.

Jantzen 2004
U. Jantzen, *Die Wasserleitung des Eupalinos. Die Funde*, Bonn, 2004.

Khamis 2010
E. Khamis, "Metal artifacts", *Ramla. Final Report on the excavations North of the White Mosque*, Qedem 50, Jeruslam, 2010, 279-285.

Kisch 1965
B. Kisch, *Scales and Weights. A Historical Outline*, New Haven and London, Yale University Press, 1965

Kürkman 2003
G. Kürkman (ed.), *Anatolian Weights and Measures*, Suna-İnan Kıraç Research Institute on Mediterranean Civilizations (trs. P.M. Işın), İstanbul, 2003.

Lightfoot 2012
C. Lightfoot, "Byzantine Weights and Related Material, 1998-2008", (ed. C. Lightfoot and E.A. Ivison), *Amorium Reports* 3, Istanbul, 2012, 255-259.

Manns 1984
F. Manns, *Some Weights of the Hellenistic, Roman and Byzantine Periods*, Jerusalem, 1984.

Mincev 2008
A. Mincev, "Early Byzantine Weights Found in North-eastern Bulgaria and Some Notes on Their Production, Distribution and Use", *Acta Varnensis* VII-2 (2008), 7-40.

Mototolea 2009
A. Mototolea, "Ponduri Romano-Bizantine din colectia Muzeului de Istorie Nationala şi Arheologie Constanta", *Pontica* XLII (2009), 663-669.

Nenci 2001
G. Nenci, "Pesi islamici di bronzo e di piombo da Entella", *Zona Archeologica: Festschrift für Hans Peter Isler zum 60. Geburstag*, Bonn, 2001, 315-317.

Ocheşeanu 1984
R. Ocheşeanu, "Ponduri romano-bizantine în colectiile din România", *Studii şi Cercetari de Numismatica* VIII (1984), 89-93.

Ocheşeanu – Cliante 1991
R. Ocheşeanu – T. Cliante, "Pond uncial descoperit la Constanta", *Pontica* XXIV (1991), 403-405.

Pink 1938
K. Pink, *Römische und byzantinische Gewichte in österreichischen Sammlungen*, Wien, 1938.

Qedar 1978
S. Qedar, *Gewichte aus 3 Jahrtausenden. Münz Zentrum* XXXII. Auktion, Köln, 1978.

Qedar 1979
S. Qedar, *Gewichte aus Drei Jahrtausenden. Teil II, Münz Zentrum* XXXVII. Auktion, Köln, 1979.

Qedar 1981
S. Qedar, *Gewichte aus Drei Jahrtausenden. Teil III, Münz Zentrum* XLV. Auktion, Köln, 1981.

Qedar 1983
S. Qedar, *Gewichte aus Drei Jahrtausenden. Teil IV, Münz Zentrum* XLIX. Auktion, Köln, 1983.

Robinson 1941
D.M. Robinson, *Excavations at Olynthus Part X: Metal and Minor Miscellaneous Finds. An Original Contribution to Greek Life*, Baltimore, London, Oxford, 1941.

Vasilopoulou 1983
V. Vasilopoulou, "Βυζαντινα σταθμια του Νομισματικου Μουσειου Αθηνων", Αρχαιολογικη Εφημερις, 122 (1983), 238-276.

Waelkens – Loots 2000
M. Waelkens – L. Loots, *Sagalassos V. Report on the Survey and Excavation Campaigns of 1996-1997*, Leuven, 2000.

Walters 1899
H.B.Walters, *Catalogue of the Bronzes. Greek, Roman, and Etruscan in the Department of Greek and Roman Antiquities, British Museum*, London, 1899.

Waldbaum 1983
J.C. Waldbaum, *Metalwork from Sardis: The Finds through 1974*, Chicago, 1983.

Weber 2009
K. Weber, *Byzantinische Münzgewichte Materialkorpus für 1-Nomisma-Gewichte*, Mass und Gewicht, Beihefte zur Zeitschrift für Metrologie, Schwelm, 2009.

Zappaterra 1999
B. Zappaterra, "Strumenti ponderali romani dal Territorio di Bondeno: pesi e bilance", *Archeologia dell'Emilia-Romagna III*, Firenze, 1999, 125-136.

NOTES ABOUT CATALOGUE

There are one hundred and sixty-one weights presented in the catalogue arranged chronologically, i.e. first ancient weights (Greek and Roman), then Byzantine weights and Islamic weights. In the catalogue lead and bronze weights are given separately. All measurements were taken twice without cleaning the weights. Some of the photographs show the items smaller while others larger than the originals. Throughout the catalogue the word "weight" is used for the subject (the material / object itself) while the "mass" is used for its weight measured in grams.

Each catalogue entry contains information in the following order:

Date / Period: The dates given are generally approximate, especially for the plain weights, which bear no denomination mark or inscription. Some forms are common and they might have been used for a long time after they had been produced. Since the weights from the archaeological contexts are rarely found it is difficult to place them in a short span of time. This is sometimes more difficult for the Islamic weights, which maintained the same form for a long time.

Geographic area / site: The possible region or site where the weight was produced and used.

Denomination: When inscribed on the item it is stated here; otherwise, when it is given theoretically in the description part below it means that the denomination is not certain.

Kind of metal: Here, only lead or bronze.

Provenance and date: Possible find place and acquisition date.

Museum's inventory number

Identification and description

Reference (cf.): It refers to the publication(s) where the similar weight(s) or information may be found; however, this part is omitted when there are no comparable items available.

HELLENISTIC AND ROMAN IMPERIAL

LEAD

1) Fourth-First centuries BC

Southern Anatolia

Lead, 60 x 44 x 18 mm, 238.80 g. Collected in Mersin and its vicinity and donated in 2010.

Museum's inv. no: 2011/1 (A) - (KP 121)

Description: Hedgehog-like in form, with hang hole at top; top is convex, bottom is flat. Its mass corresponds to tetartemorion or hemimna but it is not certain.

2) Hellenistic

Southern Anatolia

Lead, 60 x 24 mm, 350 g. Collected in Mersin and its vicinity and donated in 2010.

Museum's inv. no: 2011/2 (A) - (KP 181)

Description: Conic but flattened on the top; small break on top; on the bottom, six-armed star in relief; raised edge. Its mass corresponds to hemimna? or tritemorion, i.e. one-third of a stater; but it is not certain.

3) Hellenistic

Southern Anatolia

Lead, 25 x 28 mm, 69.45 g. Collected in Mersin and its vicinity and donated in 2010.

Museum's inv. no: 2011/3 (A) - (KP 123)

Description: Conic but flattened on the top; bottom is flat and blank.

4) Hellenistic ?

Southern Anatolia

Lead, 16 x 11 mm, 16.95 g. Collected in Mersin and its vicinity and donated in 2010.

Museum's inv. no: 2011/4 (A) - (KP 24)

Description: Conical in form; slightly flattened on the top; bottom is blank. Dating not certain. Its mass corresponds to stater or four-drachm but it is not certain and it may not be a weight.

5) Hellenistic

Southern Anatolia

Lead, 15 x 8 mm, 6.75 g. Collected in Mersin and its vicinity and donated in 2010.

Museum's inv. no: 2011/5 (A) - (KP 23)

Description: Conical in form, slightly flattened on the top; bottom is blank.

6) Hellenistic

Southern Anatolia

Lead, 15 x 6 mm, 6.15 g. Collected in Mersin and its vicinity and donated in 2010.

Museum's inv. no: 2011/6 (A) - (KP 22)

Description: Conical in form, flattened on the top. At bottom, deeply engraved X and dot in center. Denomination uncertain. It is similar in form with the preceding one thus both weights may belong to the same period.

7) Hellenistic (or later)

Southern Anatolia

Lead, 16 x 15 mm, 18.85 g. Collected in Mersin and its vicinity and donated in 2010.

Museum's inv. no: 2011/7 (A) - (KP 118)

Description: Conical with handle on top; body and bottom are blank; bottom is hollowed. Dating is not certain.

8) Hellenistic or Early Roman Imperial Period

Southern Anatolia

Lead, 38 x 46 x 9 mm, 118.70 g. Collected in Mersin and its vicinity and donated in 2010.

Museum's inv. no: 2011/8 (A) - (KP 32)

Description: Rectangular in form. Caduceus and AV in relief on the top; the bottom is blank. Its mass corresponds to tetarton, i.e. one-fourth of a mna.

9) Hellenistic or Early Roman Imperial Period

Southern Anatolia

Lead, 30 x 34 x 11 mm, 118.65 g. Collected in Mersin and its vicinity and donated in 2010.

Museum's inv. no: 2011/9 (A) - (KP 29)

Description: Square in form; both faces are blank. Worn. Its mass corresponds to tetarton, i.e. one-fourth of a mna but it is not certain.

10) Hellenistic or Early Roman Imperial Period

Southern Anatolia

Lead, 36 x 36 x 5 mm, 58.75 g. Collected in Mersin and its vicinity and donated in 2010.

Museum's inv. no: 2011/10 (A) - (KP 31)

Description: Square in form; scratched Σ or M within square border on one face, corners folded; scratched diagonal line within scratched square border on the other face. Its mass corresponds to hemisyhemitriton, i.e. one-twelfth of a stater (stater of 705 g) or one-sixteenth of a stater (hemisyhemitetarton) and as mna value, its mass corresponds to ogdoon, i.e. one-eighth of a mna but these units are not certain.

See the uninscribed square/rectangular lead weights from Olynthus see Robinson 1941, no. 2421 ff.

11) Hellenistic or Early Roman Imperial Period

Southern Anatolia

Lead, 28 x 28 x 8 mm, 48.85 g. Collected in Mersin and its vicinity and donated in 2010.

Museum's inv. no: 2011/11 (A) - (KP 30)

Description: Square in form formed by folding a rectangular plaque. Faces are blank. Its mass corresponds to two or three stater but it is not certain.

12) Hellenistic or Early Roman Imperial Period
Southern Anatolia
Lead, 23 x 25 x 8 mm, 31.20 g. Collected in Mersin and its vicinity and donated in 2010.
Museum's inv. no: 2011/12 (A) - (KP 39)
Description: Square with a slightly convex top; top and bottom is blank; worn. Its mass corresponds to distater but it is not certain.

13) Hellenistic or Early Roman Imperial Period
Southern Anatolia
Lead, 22 x 26 x 7 mm, 31.00 g. Collected in Mersin and its vicinity and donated in 2010.
Museum's inv. no: 2011/13 (A) - (KP 41)
Description: Square in form. Both faces are blank. One face is slightly concave. Its mass corresponds to distater but it is not certain.

14) Hellenistic or Early Roman Imperial Period
Southern Anatolia
Lead, 25 x 25 x 6 mm, 28.05 g. Collected in Mersin and its vicinity and donated in 2010.
Museum's inv. no: 2011/14 (A) - (KP 34)
Description: Square in form. Both faces are blank. Its mass corresponds to distater but it is not certain.

15) Hellenistic or Early Roman Imperial Period
Southern Anatolia
Lead, 17 x 22 x 7 mm, 26.05 g. Collected in Mersin and its vicinity and donated in 2010.
Museum's inv. no: 2011/15 (A) - (KP 48)
Description: Rectangular in form, brick shaped. Both faces are blank. Its mass corresponds to six-drachm but it is not certain.

16) Hellenistic or Early Roman Imperial Period
Southern Anatolia
Lead, 21 x 22 x 6 mm, 23.05 g. Collected in Mersin and its vicinity and donated in 2010.
Museum's inv. no: 2011/16 (A) - (KP 33)
Description: Square in form. Both faces are blank. Its mass corresponds to five-drachm but it is not certain.

17) Hellenistic or Roman Imperial Period

Southern Anatolia

Lead, 17 x 20 x 8 mm, 22.00 g. Collected in Mersin and its vicinity and donated in 2010.

Museum's inv. no: 2011/17 (A) - (KP 47)

Description: Square in form; both faces are blank. Worn and small break on one edge. Its mass corresponds to five-drachm but it is not certain.

For the uninscribed plain lead weights of small units see also Robinson 1941, nos. 2445-2448, 2457-2459; Mototolea 2009, Pl. II, nos. 2- 5; Ocheşeanu-Cliante 1991, Fig. 1.

18) Hellenistic or Roman Imperial Period

Southern Anatolia

Lead, 22 x 24 x 4 mm, 17.40 g. Collected in Mersin and its vicinity and donated in 2010. Its mass corresponds to tetradrachm but it is not certain.

Museum's inv. no: 2011/18 (A) - (KP 35)

Description: Square in form; both faces are blank. Its mass corresponds to stater (or tetradrachm) but it is not certain.

19) Hellenistic or Roman Imperial Period

Southern Anatolia

Lead, 16 x 19 x 6 mm, 17.20 g. Collected in Mersin and its vicinity and donated in 2010.

Museum's inv. no: 2011/19 (A) - (KP 46)

Description: Rectangular in form, brick shaped; faces are blank but one face is worn and there is a small break. Its mass may correspond to stater (tetradrachm) but it is not certain.

20) Hellenistic or Roman Imperial Period

Southern Anatolia

Lead, 18 x 18 x 5 mm, 15.95 g. Collected in Mersin and its vicinity and donated in 2010.

Museum's inv. no: 2011/20 (A) - (KP 43)

Description: Square in form; both faces are blank but scratched irregular line on one face. Its mass corresponds to stater (or tetradrachm) but it is not certain.

21) Hellenistic or Roman Imperial Period

Southern Anatolia

Lead, 17 x 21 x 5 mm, 15.40 g. Collected in Mersin and its vicinity and donated in 2010.

Museum's inv. no: 2011/21 (A) - (KP 44)

Description: Rectangular in form, brick shaped; both faces are blank. Its mass corresponds to stater (or tetradrachm) but it is not certain.

22) Hellenistic or Roman Imperial Period

Southern Anatolia

Lead, 21 x 21 x 3 mm, 14.85 g. Collected in Mersin and its vicinity and donated in 2010.

Museum's inv. no: 2011/22 (A) - (KP 40)

Description: Square in form; both faces are blank. One face is slightly concave. Its mass corresponds to stater (or tetradrachm) but it is not certain.

23) Hellenistic or Roman Imperial Period

Southern Anatolia

Lead, 17 x 18 x 5 mm, 14.50 g. Collected in Mersin and its vicinity and donated in 2010.

Museum's inv. no: 2011/23 (A) - (KP 45)

Description: Square in form; both faces are blank. Its mass corresponds to stater (or tetradrachm) but it is not certain.

24) Hellenistic or Roman Imperial Period

Southern Anatolia

Lead, 16 x 16 x 6 mm, 13.70 g. Collected in Mersin and its vicinity and donated in 2010.

Museum's inv. no: 2011/24 (A) - (KP 50)

Description: Square in form; faces are blank. Worn. Its mass may correspond to three-drachm or half uncia but it is not certain.

25) Roman Imperial Period

Southern Anatolia

Lead, 20 x 20 x 6 mm, 11.65 g. Collected in Mersin and its vicinity and donated in 2010.

Museum's inv. no: 2011/25 (A) - (KP 37)

Description: Square in form. Engraved B-like pattern on one face, other face is blank. Cut on edge. If the mark on the face refers to its unit, it may be two-nomismata/solidi but the mass is high for this unit.

26) Hellenistic or Roman Imperial Period

Southern Anatolia

Lead, 18 x 19 x 3 mm, 10.30 g. Collected in Mersin and its vicinity and donated in 2010.

Museum's inv. no: 2011/26 (A) - (KP 38)

Description: Square in form; faces are blank.

27) Hellenistic or Roman Imperial Period

Southern Anatolia

Lead, 15 x 15 x 5 mm, 10.25 g. Collected in Mersin and its vicinity and donated in 2010.

Museum's inv. no: 2011/27 (A) - (KP 49)

Description: Square in form; faces are blank.

28) Hellenistic or Roman Imperial Period

Southern Anatolia

Lead, 12 x 13 x 5 mm, 6.70 g. Collected in Mersin and its vicinity and donated in 2010.

Museum's inv. no: 2011/28 (A) - (KP 52)

Description: Square in form; both faces are blank. Its mass corresponds to two-drachm, but not certain.

29) Hellenistic or Roman Imperial Period

Southern Anatolia

Lead, 12 x 12 x 3 mm, 4.20 g. Collected in Mersin and its vicinity and donated in 2010.

Museum's inv. no: 2011/29 (A) - (KP 53)

Description: Square in form; both faces are blank. Its mass corresponds to drachm. Denomination and dating is uncertain.

30) Roman Imperial Period

Southern Anatolia

One Uncia?

Lead, 21 x 21 x 8 mm, 36.10 g. Collected in Mersin and its vicinity and donated in 2010.

Museum's inv. no: 2011/30 (A) -(KP 36)

Description: Square in form. On the top, engraved with the denominational mark A, inlaid with lead?; bottom is blank. "A" may indicate the unit "one", *i.e.* uncia; but the mass is too high for one uncia. Denomination is not certain.

31) Hellenistic or Roman Imperial Period

Southern Anatolia

Lead, 19 x 20 x 11 mm, 41.05 g. Collected in Mersin and its vicinity and donated in 2010.

Museum's inv. no: 2011/31 (A) - (KP 42)

Description: Square prism in form. Both faces are blank. It is difficult to attribute a unit for the following weights since they are plain and their masses are not standard.

32) Hellenistic or Roman Imperial Period

Southern Anatolia

Lead, 13 x 15 x 15 mm, 27.95 g. Collected in Mersin and its vicinity and donated in 2010.

Museum's inv. no: 2011/32 (A) - (KP 56)

Description: Square cube in form. Five punched dots on one facet; irregularly placed punched dots on the other facets.

33) Hellenistic or Roman Imperial Period

Southern Anatolia

Lead, 15 x 15 x 11 mm, 26.65 g. Collected in Mersin and its vicinity and donated in 2010.

Museum's inv. no: 2011/33 (A) - (KP 55)

Description: Square prism in form. Faces are blank.

34) Hellenistic- Roman Imperial Period

Southern Anatolia

Lead, 10 x 18 x 8 mm, 15.25 g. Collected in Mersin and its vicinity and donated in 2010.

Museum's inv. no: 2011/34 (A) - (KP 57)

Description: Rectangular prism or brick-like in form. Faces are blank. Its mass may correspond to stater (or tetradrachm) but it is not certain.

35) Hellenistic or Roman Imperial Period

Southern Anatolia

Lead, 11 x 10 x 7 mm, 8.15 g. Collected in Mersin and its vicinity and donated in 2010.

Museum's inv. no: 2011/35 (A) - (KP 54)

Description: Square prism in form. Faces are blank. Its mass corresponds to didrachm.

36) Roman Imperial Period (or Hellenistic?)

Southern Anatolia

Lead, 33 x 33 x 10 mm, 76.70 g. Collected in Mersin and its vicinity and donated in 2010.

Museum's inv. no: 2011/36 (A) - (KP 2)

Description: Shell in form, bottom is flat. Weights in form of shell are rare when compared with other common forms and it is difficult to date them. The mass of this piece may correspond to four-stater in Greek system or three- uncia in Roman (libra) system. Shell weights from Olynthus may be dated to the fourth century BC (then four-stater?) and pseudo-shell weights from Bondeno and Modena (both in Northern Italy) are dated to the Roman Imperial Period. Since forms of these four examples are similar to those found in Bondeno and Modena they may be dated to the Roman Imperial Period. Some scholars believe that shells of lead and of bronze were not used as scale weights. Thirtyone shells of this kind from Modena and eighty from Caesarea Maritima range in fact in a quite different dimensions and masses supporting that they may not be weights. On the other hand, the examples which have a concave bottoms filled with lead may indicate that some were used as weights. But in general we may conclude that shells of lead were not weights.

Cf. Robinson 1941, nos. 2462-2463; Crowfoot-Crowfoot-Kenyon 1957, no. 12; Manns 1984, p. 22, nos. 37-38; Zappaterra 1999, p. 134, no. 10; Corti-Pallante-Tarpini 2001, fig. 206; Hendin 2007, nos. 271-276; Hadat 2007, fig. 2; Holland 2009, pp. 59-64, fig. 7.

37) Late Hellenistic or Roman Imperial Period

Southern Anatolia

Lead, 30 x 33 x 10 mm, 50.00 g. Collected in Mersin and its vicinity and donated in 2010.

Museum's inv. no: KP 2011/37 (A) - (KP 1)

Description: Shell in form, bottom is flat. Its mass may correspond to two-uncia in Roman system or to three-stater in Greek system.

See comments and references above.

38) Late Hellenistic or Roman Imperial Period

Southern Anatolia

Lead, 25 x 28 x 11 mm, 38.10 g. Collected in Mersin and its vicinity and donated in 2010.

Museum's inv. no: KP 2011/38 (A) - (KP 4)

Description: Shell in form, the bottom is flat. Its mass may correspond to one and half uncia in Roman system or two-stater in Greek system.

See comments and references at cat. no. 36.

No. 36

No. 37

No. 38

No. 39

No. 40

39) Roman Imperial Period

Southern Anatolia

Lead, 25 x 26 x 9 mm, 22.35 g. Collected in Mersin and its vicinity and donated in 2010.

Museum's inv. no: 2011/39 (A) - (KP 3)

Description: Shell in form, the bottom is flat. Its mass may correspond to one uncia in Roman system or one stater in Greek system although its mass is low for uncia and high for stater.

See comments and references at cat. no. 36.

40) Roman Imperial Period

Southern Anatolia

Litra (Libra)

Lead, 93 x 58 x 9 mm (including tab), 350 g. Collected in Mersin and its vicinity and donated in 2010.

Museum's inv. no: 2011/40 (A) - (KP 122)

Description: Rectangular cutting board in form, with tab. On top, inscribed ΛΙΤΡΑ and *caduceus* both in relief, all in double border of dots (dots in inner row are large while the dots in outer row are small). Since Hermes is also the patron god of trade and merchants (Hermes Agoraios), it is normal that his symbol *caduceus* to be depicted on a weight. There are many weights which bear *caduceus* on them in the private and public collections. The bottom are covered with a crosshatching pattern in relief. Some parts are crushed.

Cf. Gorny und Mosch 75, 9 March 2009, # 343 (hemilitron, 156.19 g).

41) Hellenistic?

Southern Anatolia

Lead, 17 x 5 mm, 10.15 g. Collected in Mersin and its vicinity and donated in 2010.

Museum's inv. no: 2011/41 (A) - (KP 21)

Description: Irregular hexagonal; both faces are blank or illegible; worn, some parts crushed. Dating not certain and may not be a weight.

BRONZE

42) Hellenistic - Roman Imperial Period

Southern Anatolia

Bronze, 40 x 15 x 5 mm, 16.55 g. Collected in Mersin and its vicinity and donated in 2010.

Museum's inv. no: 2011/42 (A) - KP 119

Description: Pelta in form, both sides are flat but on one side, at center, a deep scrape. Its mass may correspond to stater or four-drachm. It may not be a weight but appliqué or phalera.

Cf. Davidson 1952, no. 539; Boucher 1971, nos. 858-60.

43) Hellenistic - Roman Imperial Period

Southern Anatolia

Bronze, 34 x 11 x 3 mm, 3.80 g. Collected in Mersin and its vicinity and donated in 2010.

Museum's inv. no: 2011/43 (A) - KP 120

Description: Pelta in form, both sides are flat. It is nearly a quarter in unit when it is compared with the preceding one. Its mass may correspond to one drachm. It may not be a weight but appliqué or phalera.

Cf. Davidson 1952, no. 539; Boucher 1971, nos. 858-60.

44) Roman Imperial Period (or Hellenistic?)

Southern Anatolia

Bronze, 19 x 20 x 7 mm, 9.75 g. Collected in Mersin and its vicinity and donated in 2010.

Museum's inv. no: 2011/44 (A) - KP 5

Description: Shell in form, the bottom is concave. If there was a lead inside the hollow bottom its original mass should be higher. It may be a half-uncia in Roman system or didrachm in Greek system, but possibly not a weight.

See the references and comment at cat. no. 36.

45) Hellenistic – Early Roman Imperial Period

Southern Anatolia

Bronze, 27 x 27 x 15 mm, 82.70 g. Collected in Mersin and its vicinity and donated in 2010.

Museum's inv. no: 2011/45 (A) - KP 64

Description: Square prism in form; both faces are blank. It may correspond to hektemorion (1/6 mna) but it is not certain.

46) Late Roman Imperial (AD 3rd - 4th centuries)

Southern Anatolia

Two-uncia

Bronze, 23 x 18 mm, 52.60 g. Collected in Mersin and its vicinity and donated in 2010.

Museum's inv. no: 2011/46 (A) - KP 149

Description: The weight is a flattened sphere doubly truncated in form; centring points on the top and bottom; engraved on the top with the denominational mark Ɣ B; inlaid with silver; bottom and body are blank; patinated.

Cf. Walters 1899, no. 3029; Davidson 1952, no.1586; Dürr 1964, no. 269 and 307; Qedar 1978, no. 84; Manns 1984, p. 31, no. 14; Qedar 1981, no. 108; Binsfeld 1990, 27-28; Bendall 1996, no. 31; Hendin 2007, no. 331; *Byzanz. Licht aus dem Osten,* no. III.33; Kürkman 2003, no. 59; *Die Welt von Byzanz 2004,* no. 792; Minchev 2008, p. 26, 111, fi. 70.

47) Late Roman Imperial (AD 3rd - 4th centuries)

Southern Anatolia

Uncia

Bronze, 18 x 14 mm, 26.15 g. Collected in Mersin and its vicinity and donated in 2010.

Museum's inv. no: KP 2011/47 (A) - 145

Description: The weight is a flattened sphere doubly truncated in form; centring points on the top and bottom; engraved on the top with the denominational mark Ɣ A, inlaid with lead?; bottom and body are blank; patinated.

Cf. Walters 1899, no. 3030; Pink 1938, no. 46; Davidson 1952, no. 1585; Dürr 1964, no. 263; Qedar 1981, nos. 108; Waldbaum 1983, no. 475; *Byzanz. Licht aus dem Osten,* no. III.34; Binsfeld 1990, 29-30; Bendall 1996, no. 33; Kürkman 2003, no. 58; *Die Welt von Byzanz 2004,* no. 793; Hendin 2007, no. 332; Minchev 2008, p. 24, nos. 100-101, figs. 62-63.

48) Late Roman Imperial (AD 3rd-4th centuries)

Southern Anatolia

Bronze, 17 x 15 mm, 24.95 g. Collected in Mersin and its vicinity and donated in 2010.

Museum's inv. no: 2011/48 (A) - (KP 144)

Description: The weight is a flattened sphere doubly truncated in form; centring points on the top and bottom; top, bottom and body are blank or illegible. Corroded and pitted. The attribution to a Roman weight is not certain since it doesn't bear any mark on it; it is difficult to determine whether it is a Roman uncia or an Islamic 10 dirham weight. But its spherical form without carinate on the profile body and centrings points give an impression that it may be a Byzantine uncia. A weight similar in form (perhaps two-nomismata) was found at the excavations at Corinth.

Cf. Davidson 1952, no. 1584 (Roman); Isler 1998, no. 25 (Islamic).

49) Late Roman Imperial -Early Byzantine

Southern Anatolia

1 sicilus?

Bronze, 12 x 10 mm, 6.65 g. Collected in Mersin and its vicinity and donated in 2010.

Museum's inv. no: 2011/49 (A) - (KP 27)

Description: Sphere flattened only on one side. Too worn, corroded and pitted. Its mass corresponds to sicilus (quarter uncia) but dating and unit are not certain; it may be Islamic, too.

BYZANTINE

LEAD

50) Early Byzantine (AD 5th-7th centuries)

Southern Anatolia

Lead, 15 x 16 x 7 mm, 14.45 g. Collected in Mersin and its vicinity and donated in 2010.

Museum's inv. no: 2011/50 (A) - (KP 51)

Description: Square in form. Deep engraved + (or cross) on both faces. Its mass may correspond to half-uncia. If we accept the cross as + then the weight may be dated to the Hellenistic period (hemistater?). Dating and denomination are not certain.

BRONZE

51) Early Byzantine (AD 6th-7th centuries)

Southern Anatolia

Four-nomismata

Bronze, 24 x 7 mm, 17.95 g. Collected in Mersin and its vicinity and donated in 2010.

Museum's inv. no: 2011/51 (A) - (KP 6)

Description: Discoid, with raised rims and centring points on the top and bottom; convex double-grooved profile. Both faces are illegible. On the top there should be a denomination mark (N Δ) but it is illegible or never existed. Small marks on one side. Pitted and corroded.

Cf. Waldbaum 1983, no. 478; Jantzen 2004, Taf. 38, no. 1250 (51.63 g).

52) Early Byzantine (AD 6th-7th centuries)

Southern Anatolia

Three-nomismata

Bronze, 20 x 6 mm, 13.45 g. Collected in Mersin and its vicinity and donated in 2010.

Museum's inv. no: 2011/52 (A) - (KP 7)

Description: Discoid with raised rims and centring points on the top and bottom; convex double-grooved profile; punched on the top with denominational mark N Γ ; above and below, punched geometric wave pattern. On bottom, three punched dots. Crashed on one side; patinated.

Cf. Draganov 1994, no. 2; Waelkens-Loots 2000, fig. 103 (six nomismata); Minchev 2008, p. 21, no. 86, fig. 50.

Cat. no. 56

Cat. no. 67

Cat. no. 66

53) Early Byzantine

Southern Anatolia

Bronze, 17 x 2 mm, 4.15 g. Collected in Mersin and its vicinity and donated in 2010.

Museum's inv. no: 2011/53 (A) - (KP 18)

Description: Discoid; engraved a star with eight-arms on one face; other face is blank. Its mass corresponds to one nomisma.

Cf. Hendin 2007, no. 368; Weber 2009, Taf. 18, no. 24.2.

54) Early Byzantine (AD 5th-7th centuries)

Southern Anatolia

Bronze, 17 x 2 mm, 3.70 g. Collected in Mersin and its vicinity and donated in 2010.

Museum's inv. no: 2011/54 (A) - (KP 15)

Description: Discoid; concentric circles encircling the centring point on the top and engraved X on the bottom. Its mass corresponds to one nomisma but the mass is low for nomisma. Worn, corroded.

Cf. Kürkman 2003, no. 44; Jantzen 2004, no. 1235; Weber 2009, Abbildung no. 25 = no. 255, no. 177.

55) Early Byzantine (AD 5th-7th centuries)

Southern Anatolia

Bronze, 13 x 6 mm, 3.45 g. Collected in Mersin and its vicinity and donated in 2010.

Museum's inv. no: 2011/55 (A) - (KP 8)

Description: Discoid with a deep groove near the edge encircling the prominent centring point; too worn, corroded and pitted. Its mass may correspond to nomisma but the mass is low for nomisma.

Cf. Weber 2009, no. 319 (same design on both sides).

56) Byzantine (AD 6th-7th centuries)

Southern Anatolia

Two-uncia

Bronze, 25 x 27 x 10 mm, 52.05 g. Collected in Mersin and its vicinity and donated in 2010.

Museum's inv. no: 2011/56 (A) - (KP 63)

Description: Square in form. Engraved on the top with the denominational mark Γο B and cross above with punched dots at the extremities; originally not inlaid; bottom is blank; patinated.

Cf. Dürr 1964, no. 59.

57) Byzantine (6th-7th centuries)

Southern Anatolia

Bronze, 20 x 21 x 5 mm, 18.15 g. Collected in Mersin and its vicinity and donated in 2010.

Museum's inv. no: 2011/57 (A) - (KP 66)

Description: Square in form. Both faces and profile are blank. Patinated and corroded. Its mass correspond to four-nomismata.

58) Byzantine (AD 6th-7th centuries)

Southern Anatolia

Two-nomismata

Bronze, 16 x 16 x 4 mm, 9.25 g. Collected in Mersin and its vicinity and donated in 2010.

Museum's inv. no: 2011/58 (A) - (KP 68)

Description: Square in form. On the top, engraved illegible denominational mark (N B?) within wreath, cross illegible or no cross; originally inlaid but inlay is missing although traces of corroded inlay are still visible; bottom and profile is blank and flat. Patinated and corroded.

Cf. Vasilopoulou 1985, no. 42; Bendall 1996, no. 102.

59) Byzantine (AD 6th-7th centuries)

Southern Anatolia

Two-nomismata

Bronze, 17 x 17 x 3 mm, 9.20 g. Collected in Mersin and its vicinity and donated in 2010.

Museum's inv. no: 2011/59 (A) - (KP 67)

Description: Square in form. Engraved on the top with the denominational mark N B, cross illegible or no cross; originally inlaid with lead, wreath illegible or no wreath; bottom and profile is blank and flat.

Cf. Vasilopoulou 1985, no. 42; Bendall 1996, no. 102.

60) Early Byzantine (AD 5th-7th centuries)

Southern Anatolia

Bronze, 27 x 27 x 2 mm, 8.60 g. Collected in Mersin and its vicinity and donated in 2010.

Museum's inv. no: 2011/60 (A) - (KP 65)

Description: Square in form; both faces are blank (or illegible). Patinated, corroded and pitted. Its mass may correspond to two nomismata.

61) Early Byzantine (AD 5th-7th centuries)

Southern Anatolia

Bronze, 13 x 14 x 3 mm, 4.55 g. Collected in Mersin and its vicinity and donated in 2010.

Museum's inv. no: 2011/61 (A) - (KP 77)

Description: Square in form. Both faces are blank. Patinated and corroded. Its mass corresponds one nomisma.

62) Early Byzantine (AD 4th-6th centuries)

Southern Anatolia

Nomisma

Bronze, 11 x 12 x 3 mm, 4.50 g. Collected in Mersin and its vicinity and donated in 2010.

Museum's inv. no: 2011/62 (A) - (KP 89)

Description: Square in form. Engraved N on the top, originally not inlaid; bottom and profile is flat and blank.

Cf. Delattre 1918, p. 182, nos. 44-46; Dürr 1964, nos. 156 ff.; Kürkman 2003, no. 34; Vasilopoulou 1985, nos. 46, 49; Hendin 2007, no. 356; Weber 2009, no. 28, 37.

63) Early Byzantine (AD 4th-6th centuries)

Southern Anatolia

Nomisma

Bronze, 13 x 15 x 3 mm, 4.45 g. Collected in Mersin and its vicinity and donated in 2010.

Museum's inv. no: 2011/63 (A) - (KP 78)

Description: Square in form. Punched N on the top; bottom is blank; flat profile. Patinated and corroded.

Cf. Minchev 2008, no. p. 14, no. 41, fig. 22; Weber 2009, no. 54.

64) Early Byzantine (AD 4th-6th centuries)

Southern Anatolia

Bronze, 13 x 13 x 3 mm, 4.40 g. Collected in Mersin and its vicinity and donated in 2010.

Museum's inv. no: 2011/64 (A) - (KP 81)

Description: Square in form. On the top, engraved X within square border, originally not inlaid; bottom is blank. Its mass corresponds to nomisma.

Cf. Dürr 1964, nos. 283, 291, 293, 294 and 295; Weber 2009, 374.

65) Early Byzantine (AD 4th-6th centuries)

Southern Anatolia

Nomisma

Bronze, 12 x 13 x 3 mm, 4.40 g. Collected in Mersin and its vicinity and donated in 2010.

Museum's inv. no: 2011/65 (A) - (KP 83)

Description: Square in form. Engraved on the top with the denominational mark N̊; originally not inlaid. Bottom is blank; flat profile; patinated.

Cf. Ocheşeanu 1984, no. 17; Minchev 2008, p. 14, no. 36, fig. 17; Weber 2009, no. 205; Lightfoot 2012, p. 257, no. 8.

66) Early Byzantine (AD 4th-6th centuries)

Southern Anatolia

Nomisma

Bronze, 12 x 13 x 3 mm, 4.40 g. Collected in Mersin and its vicinity and donated in 2010.

Museum's inv. no: 2011/66 (A) - (KP 85)

Description: Square in form. Engraved on the top with the denominational mark N̊; originally not inlaid. Bottom is blank; flat profile; patinated.

Cf. Dürr 1964, no. 205.

67) Early Byzantine (AD 4th-6th centuries)

Southern Anatolia

Nomisma

Bronze, 11 x 11 x 4 mm, 4.35 g. Collected in Mersin and its vicinity and donated in 2010.

Museum's inv. no: 2011/67 (A) - KP 92

Description: Square in form. Punched on the top with the denominational mark N; originally not inlaid. Bottom is blank; flat profil; patinated.

68) Early Byzantine (AD 4th-6th centuries)

Southern Anatolia

Nomisma

Bronze, 12 x 13 x 3 mm, 4.25 g. Collected in Mersin and its vicinity and donated in 2010.

Museum's inv. no: 2011/68 (A) - (KP 75)

Description: Square in form. Engraved on the top with the denominational mark N̊; and engraved X and punched rosettes between the arms on the bottom; flat profile; patinated.

Cf. Weber 2009, 365 ff.

69) Early Byzantine (AD 4th-6th centuries)

Southern Anatolia

Bronze, 14 x 14 x 3 mm, 4.15 g. Collected in Mersin and its vicinity and donated in 2010.

Museum's inv. no: 2011/69 (A) - (KP 72)

Description: Square in form. Both faces are blank or illegible; flat profile. Patinated, corroded and pitted. Its mass corresponds to nomisma.

70) Early Byzantine (AD 4th-6th centuries)

Southern Anatolia

Bronze, 12 x 13 x 3 mm, 4.05 g. Collected in Mersin and its vicinity and donated in 2010.

Museum's inv. no: 2011/70 (A) - (KP 88)

Description: Square in form. On the top, engraved X terminating with motif of three punched dots; there are dots near edge forming a border; bottom and profile are blank. Patinated and corroded. Its mass corresponds to nomisma.

Minchev 2008, p. 17, no. 61, fig. 34.

71) Early Byzantine (AD 4th-6th centuries)

Southern Anatolia

Nomisma

Bronze, 11 x 12 x 4 mm., 3.95 g. Collected in Mersin and its vicinity and donated in 2010.

Museum's inv. no: 2011/71 (A) - (KP 95))

Description: Square in form. Engraved on the top with the denominational mark N, punched dots on the extremities of the letter and two dots between legs; bottom and profile is blank and flat.

Cf. Weber 2009, 27 and 38.

72) Early Byzantine (AD 4th-6th centuries)

Southern Anatolia

Nomisma

Bronze, 13 x 13 x 3 mm, 3.90 g. Collected in Mersin and its vicinity and donated in 2010.

Museum's inv. no: 2011/72 (A) - KP 74

Description: Square in form. Engraved on the top with the denominational mark N, punched dots on the extremities of the letter and two dots between legs; bottom is blank and flat profile.

Cf. Binsfeld 1990, no. 44; Kürkman 2003, no. 36; Minchev 2008, p. 14, no. 37, fig. 18; Weber 2009, *nos.* 27 and 38.

73) Early Byzantine (AD 4th-6th centuries)

Southern Anatolia

Nomisma

Bronze, 14 x 15 x 2 mm, 3.70 g. Collected in Mersin and its vicinity and donated in 2010.

Museum's inv. no: 2011/73 (A) - KP 80

Description: Square in form. Engraved on the top with the denominational mark N and a motif of three punched dots, but all hardly visible; bottom is blank; flat profile. Patinated, corroded and pitted.

74) Early Byzantine (AD 4th-6th centuries)

Southern Anatolia

Bronze, 15 x 15 x 2 mm, 3.30 g. Collected in Mersin and its vicinity and donated in 2010.

Museum's inv. no: 2011/74 (A) - KP 71

Description: Square in form. Both faces are blank or illegible; flat profile. Patinated and corroded. Its mass corresponds to nomisma? (or rather three-scruples).

75) Early Byzantine (AD 4th-6th centuries)

Southern Anatolia

Bronze, 10 x 10 x 2 mm, 2.30 g. Collected in Mersin and its vicinity and donated in 2010.

Museum's inv. no: 2011/75 (A) - KP 103

Description: Square in form. On the top, punched dots along the edge; bottom is blank; flat profile. Patinated and corroded. Its mass corresponds to twelve-siliquae (or half nomisma).

76) Early Byzantine (AD 4th-6th centuries)

Southern Anatolia

Twelve siliquae (or half nomisma)

Bronze, 9 x 10 x 2 mm., 2.15 g. Collected in Mersin and its vicinity and donated in 2010.

Museum's inv. no: : 2011/76 (A) - (KP 109)

Description: Square in form. Engraved on the top with the denominational mark IB; originally not inlaid. Bottom is blank; flat profile. Patinated.

Cf. Mincev 2008, p. 12, n0. 24-25, fig. 11-12; Weber 2009, no. 51.1

77) Early Byzantine (AD 4th-6th centuries)

Southern Anatolia

Bronze, 9 x 9 x 3 mm, 2.15 g. Collected in Mersin and its vicinity and donated in 2010.

Museum's inv. no: 2011/77 (A) - (KP 113)

Description: Square in form. Both faces are blank; flat profile. Patinated and corroded. Its mass corresponds to twelve-siliquae (or half nomisma). Attribution and dating are not certain.

78) Early Byzantine? (AD 4th-6th centuries?)

Southern Anatolia

Bronze, 8 x 8 x 3 mm, 1.50 g. Collected in Mersin and its vicinity and donated in 2010.

Museum's inv. no: 2011/78 (A) - (KP 114)

Description: Square in form (nearly cubic). Both faces are blank; flat profile. Patinated and corroded. Its mass corresponds to eight-siliquae. Dating and denomination are not certain.

79) Early Byzantine (AD 4th-6th centuries?)

Southern Anatolia

Bronze, 8 x 8 x 3 mm, 1.40 g. Collected in Mersin and its vicinity and donated in 2010.

Museum's inv. no: 2011/79 (A) - (KP 116)

Description: Square in form. On one face, stamped cross within circular border and illegible letters around. On the other face, three punched dots. Patinated and corroded. Its mass corresponds to eight-siliquae.

80) Early Byzantine? (AD 4th-6th centuries?)

Southern Anatolia

Bronze, 9 x 9 x 3 mm, 1.00 g. Collected in Mersin and its vicinity and donated in 2010.

Museum's inv. no: 2011/80 (A) - (KP 117)

Description: Square in form. Both faces are blank. Low mass due to the break at edge. Since the faces are blank (or illegible) and much corroded attribution and dating are not certain. Its mass corresponds to five-siliquae or low eight-siliquae?

81) Byzantine?

Southern Anatolia

Bronze, 8 x 10 x 2 mm, 0.60 g. Collected in Mersin and its vicinity and donated in 2010.

Museum's inv. no: 2011/81 (A) - (KP 110)

Description: Square in form formed by being folded of a rectangular plaque. On the top, engraved X and punched circular dots; bottom is blank. Patinated and corroded. Its mass may correspond to three-siliquae. Attribution, dating and denomination are not certain. May not be a weight.

82) Uncertain (Hellenistic, Roman Imperial, Byzantine?)

Southern Anatolia

Bronze, 13 x 17 x 6 mm, 9.50 g. Collected in Mersin and its vicinity and donated in 2010.

Museum's inv. no: 2011/82 (A) - (KP 70)

Description: Rectangular in form. Faces are blank. Patinated and corroded.

ISLAMIC

(All dates are AD)

LEAD

83) Islamic (9th-13th centuries)

Southern Anatolia

Lead, 20 x 13 mm, 32.10 g. Collected in Mersin and its vicinity and donated in 2010.

Museum's inv. no: 2011/83 (A) - (KP 143)

Description: The weight is a flattened sphere doubly truncated in form (slightly barrel-like); carinated profile; top, bottom and body are blank or illegible. Corroded. Its mass corresponds to twelve-dirham.

Cf. Isler 1998, no. 12.

BRONZE

84) Islamic or Byzantine (9th-13th centuries?)

Southern Anatolia

Bronze, 18 x 9 mm, 18.10 g. Collected in Mersin and its vicinity and donated in 2010.

Museum's inv. no: 2011/84 (A) - (KP 12)

Description: Discoid, both faces are blank; breaks on edges; much corroded. Since both sides are blank and shows no characteristics it is difficult to date. It may be a four-mithqal weight; its mass corresponds also to Byzantine four-nomismata.

85) Islamic? (9th-13th centuries?)

Southern Anatolia

Bronze, 16 x 6 mm, 8.40 g. Collected in Mersin and its vicinity and donated in 2010.

Museum's inv. no: 2011/85 (A) - (KP 11)

Description: Discoid, both faces are blank; slight carinated profile. Patinated and corroded. Since both faces are blank and shows no characteristic, dating is difficult. The mass corresponds to two-mithqal and also to three-dirham.

86) Islamic (9th-11th centuries)

Southern Anatolia

Two-Mithqal or three-dirham

Bronze, 16 x 5 mm, 8.10 g. Collected in Mersin and its vicinity and donated in 2010.

Museum's inv. no: 2011/86 (A) - (KP 17)

Description: Discoid in form. Two bird's eye patterns on both faces. Patinated and corroded.

Cf. Qedar 1979, no. 4207.

87) Islamic (9th-11th centuries)

Southern Anatolia

Two-Mithqal or three dirham

Bronze, 15 x 5 mm, 8.10 g. Collected in Mersin and its vicinity and donated in 2010.

Museum's inv. no: 2011/87 (A) - (KP 9)

Description: Discoid with one bird's eye pattern (in fact a circular groove encircling the centring point) on both faces. Patinated.

Cf. Qedar 1979, no. 4207.

88) Islamic (9th-11th centuries)

Southern Anatolia

Bronze, 15 x 7 mm, 7.95 g. Collected in Mersin and its vicinity and donated in 2010.

Museum's inv. no: 2011/88 (A) - (KP 10)

Description: Discoid, both faces are blank. Patinated and corroded. Since both faces are blank and flat its attribution is difficult. Its mass corresponds to three-dirham or two-mithqal, but it is not certain.

Cf. Qedar 1979, no. 4207.

89) Islamic (9th-11th centuries)

Southern Anatolia

Bronze, 14 x 7 mm, 7.65 g. Collected in Mersin and its vicinity and donated in 2010.

Museum's inv. no: 2011/89 (A) - (KP 14)

Description: Discoid with one bird's eye pattern on one face and two bird's eye patterns on the other face. Slightly carinated profile. Patinated and corroded. It may be Islamic three-dirham of 2.55 g x 3.

Cf. Qedar 1979, no. 4207.

90) Islamic? (9th-11th centuries)

Southern Anatolia

Mithqal

Bronze, 12 x 6 mm., 4.10 g. Collected in Mersin and its vicinity and donated in 2010.

Museum's inv. no: 2011/90 (A) - (KP 19)

Description: Discoid with two bird's eye pattern on one face, other face is illegible (or blank). Its mass corresponds also to Byzantine nomisma. Patinated and corroded.

Cf. Qedar 1979, no. 4212.

91) Islamic (9th-11th centuries)

Southern Anatolia

Mithqal

Bronze, 12 x 5 mm, 4.05 g. Collected in Mersin and its vicinity and donated in 2010.

Museum's inv. no: 2011/91 (A) - (KP 13)

Description: Discoid with two bird's eye patterns on both faces. Slightly carinated profile. Patinated and corroded. Its mass corresponds to mithqal.

Cf. Qedar 1979, no. 4212.

92) Islamic (9th-11th centuries)

Southern Anatolia

Mithqal

Bronze, 11 x 5 mm, 4.05 g. Collected in Mersin and its vicinity and donated in 2010.

Museum's inv. no: 2011/92 (A) 2011/1 (A) - (KP 16)

Description: Discoid with two bird's eye patterns on both faces. Slightly carinated profile. Its mass corresponds to mithqal.

Cf. Qedar 1979, no. 4212.

93) Islamic (11th-13th centuries)

Southern Anatolia

Ten-dirham

Bronze, 15 x 15 x 15 mm, 29.45 g. Collected in Mersin and its vicinity and donated in 2010.

Museum's inv. no: 2011/93 (A) - (KP 60)

Description: Square cube in form. On all faces, one central large bird's eye and four smaller bird's eyes at each corner. Corroded. Its mass corresponds to ten-dirham.

Cf. Holland 1986, no. 53-58 (different units); Hendin 2007, no. 441 (five dirham); Kürkman 2003, no. 94 (five-dirham).

94) Islamic (11th-13th centuries)

Southern Anatolia

Ten-dirham

Bronze, 15 x 15 x 15 mm, 29.05 g. Collected in Mersin and its vicinity and donated in 2010.

Museum's inv. no: 2011/94 (A) - (KP 61)

Description: Square cube in form. Double bird's eyes on each face. Its mass corresponds to ten-dirham.

Cf. Holland 1986, no. 53-58 (different units); Hendin 2007, no. 441 (five dirham); Kürkman 2003, no. 94 (five-dirham).

95) Islamic (10th-13th centuries)

Southern Anatolia

Ten-dirham

Bronze, 18 x 18 x 11 mm., 28.15 g. Collected in Mersin and its vicinity and donated in 2010.

Museum's inv. no: 2011/95 (A) - (KP 62)

Description: Square prism in form. Punched identical circular stamp containing kufic script (Ayar-ı Bekke/Mekke?) on two large faces; other faces are blank. Its mass corresponds to ten-dirham.

96) Islamic (10th-13th centuries)

Southern Anatolia

Eight-dirham

Bronze, 15 x 10 x 10 mm, 23.60 g. Collected in Mersin and its vicinity and donated in 2010.

Museum's inv. no: 2011/96 (A) - (KP 59)

Description: Square prism in form. Four bird's eyes on large faces and two bird's eye on narrow faces. It should be an eight-dirham of 2.95 g x 8; otherwise it may be a ten-dirham but its mass is low for ten-dirham.

97) Islamic (10th-13th centuries)

Southern Anatolia

Five-dirham

Bronze, 12 x 12 x 13 mm, 14.65 g. Collected in Mersin and its vicinity and donated in 2010.

Museum's inv. no: 2011/97 (A) - (KP 58)

Description: Square cube in form. One large bird's eye on all faces except one. Patinated and corroded.

Cf. Holland 1986, no. 53-58 (different units); Hendin 2007, no. 441 (five dirham); Kürkman 2003, no. 94 (five-dirham); Holland 2009, nos. 17-122 (different units).

98) Islamic (10th-13th centuries)
Southern Anatolia
Two-dirham
Bronze, 11 x 14 x 6 mm, 6.00 g. Collected in Mersin and its vicinity and donated in 2010.
Museum's inv. no: 2011/98 (A) - (KP 97)
Description: Rectangular cube (nearly) in form. All faces are blank. Patinated and corroded. Dating and denomination is not certain.

99) Islamic (10th-13th centuries)
Southern Anatolia
Two-dirham
Bronze, 14 x 15 x 4 mm, 5.90 g. Collected in Mersin and its vicinity and donated in 2010.
Museum's inv. no: 2011/99 (A) - (KP 76)
Description: Rectangular in form. Double bird's eyes on both faces; flat profile. Patinated and corroded.
Cf. Hendin 2007, no. 444.; Kürkman 2003, nos. 89-91.

100) Islamic (10th-13th centuries)
Southern Anatolia
Two-dirham
Bronze, 13 x 13 x 4 mm, 5.85 g. Collected in Mersin and its vicinity and donated in 2010.
Museum's inv. no: 2011/100 (A) - (KP 73)
Description: Square in form. Double bird's eyes on both faces; thinny carinated profile. Patinated and corroded.
Cf. Hendin 2007, no. 444; Kürkman 2003, nos. 89-91.

101) Islamic (10th-13th centuries)
Southern Anatolia
Two-dirham
Bronze, 13 x 15 x 3 mm, 5.80 g. Collected in Mersin and its vicinity and donated in 2010.
Museum's inv. no: 2011/101 (A) - (KP 79)
Description: Square in form; both faces are blank (or illegible). Patinated and corroded.

102) Islamic (10th-13th centuries)
Southern Anatolia
Two-dirham
Bronze, 13 x 13 x 4 mm, 5.80 g. Collected in Mersin and its vicinity and donated in 2010.
Museum's inv. no: 2011/102 (A) - (KP 84)
Description: Square in form. Large bird's eye on both faces; slightly carinated profile. Patinated and corroded.
Cf. Hendin 2007, no. 449 (one dirham).

103) Islamic (10th-13th centuries)

Southern Anatolia

Two-dirham

Bronze, 13 x 13 x 5 mm, 5.80 g. Collected in Mersin and its vicinity and donated in 2010.

Museum's inv. no: 2011/103 (A) - (KP 86)

Description: Square in form. Double bird's eyes on both faces; flat profile. Patinated and corroded.

Cf. Hendin 2007, no. 444; Kürkman 2003, nos. 89-91.

104) Islamic (10th-13th centuries)

Southern Anatolia

Two-dirham

Bronze, 12 x 14 x 5 mm, 5.80 g. Collected in Mersin and its vicinity and donated in 2010.

Museum's inv. no: 2011/104 (A) - (KP 87)

Description: Square in form. On one face, a circular stamp including kufic script (may be "Abd 7"); on the other face, two bird's eyes. Patinated and corroded.

Cf. Kürkman 2003, no. 85 (one dirham); also here cat. nos. 104 and 115.

105) Islamic (10th-13th centuries)

Southern Anatolia

Two-dirham

Bronze, 12 x 13 x 5 mm, 5.75 g. Collected in Mersin and its vicinity and donated in 2010.

Museum's inv. no: 2011/105 (A) - (KP 94)

Description: Square in form. Double bird's eyes on both faces (bird's eyes on one face is hardly visible); patinated.

Cf. Hendin 2007, no. 444.

106) Islamic (10th-13th centuries)

Southern Anatolia

Two-dirham

Bronze, 11 x 11 x 7 mm, 5.75 g. Collected in Mersin and its vicinity and donated in 2010.

Museum's inv. no: 2011/106 (A) - (KP 99)

Description: Square in form. Both faces are blank; thinny carinated profile. Patinated and corroded.

107) Islamic (10th-13th centuries)

Southern Anatolia

Two-dirham

Bronze, 10 x 14 x 6 mm, 5.75 g. Collected in Mersin and its vicinity and donated in 2010.

Museum's inv. no: 2011/107 (A) - (KP 96)

Description: Rectangular cube (nearly) in form. Two bird's eyes and two identical stamps with kufic script (may be "Abd 7") on one face; two bird's eyes on the other face; narrow faces are blank. Small part on one corner broken away; patinated and corroded.

Cf. Qedar 1979, no. 4210 (no inscription). For similar script on one dirham weight see Kürkman 2003, no. 85 read as "Muhammed"; also here cat. nos. 104 and 115.

108) Islamic (10th-13th centuries)

Southern Anatolia

Two-dirham

Bronze, 12 x 13 x 5 mm, 5.70 g. Collected in Mersin and its vicinity and donated in 2010.

Museum's inv. no: 2011/108 (A) - (KP 93)

Description: Square in form. Double bird's eyes on both faces. Patinated and corroded.

Cf. Hendin 2007, no. 444; Holland 2009, no. 131.

109) Islamic (10th-13th centuries)

Southern Anatolia

Two-dirham

Bronze, 11 x 12 x 6 mm, 5.70 g. Collected in Mersin and its vicinity and donated in 2010.

Museum's inv. no: 2011/109 (A) - (KP 91)

Description: Thick square in form. On both faces, one large central bird's eye and four smaller bird's eyes at each corner; flat profile. Patinated and corroded.

Holland 2009, no. 130.

110) Islamic (10th-13th centuries)

Southern Anatolia

Bronze, 13 x 14 x 5 mm, 5.60 g. Collected in Mersin and its vicinity and donated in 2010.

Museum's inv. no: 2011/110 (A) - (KP 82)

Description: Square in form. Both faces are blank (or illegible). Heavily corroded. Its mass corresponds to two-dirham.

111) Islamic (10th-13th centuries)

Southern Anatolia

Two-dirham

Bronze, 11 x 12 x 3 mm, 5.40 g. Collected in Mersin and its vicinity and donated in 2010.

Museum's inv. no: 2011/111 (A) - (KP 90)

Description: Square in form. Double bird's eyes on both faces; thinny carinated on two profile. Patinated and corroded.

Cf. Hendin 2007, no. 444.

112) Islamic (10th-13th centuries)

Southern Anatolia

Dirham

Bronze, 12 x 11 x 5 mm, 3.00 g. Collected in Mersin and its vicinity and donated in 2010.

Museum's inv. no: 2011/112 (A) - (KP 105)

Description: Nearly square in form. Faces are blank or illegible. Heavily corroded.

113) Islamic (10th-13th centuries)

Southern Anatolia

Dirham

00) Bronze, 9 x 9 x 5 mm, 2.95 g. Collected in Mersin and its vicinity and donated in 2010.

Museum's inv. no: 2011/113 (A) - (KP 108)

Description: Square (nearly cube) in form. Bird's eye on each face; flat profile. *Cf.* Hendin 2007, no. 449.

114) Islamic (10th-13th centuries)

Southern Anatolia

Dirham

Bronze, 10 x 11 x 3 mm, 2.90 g. Collected in Mersin and its vicinity and donated in 2010.

Museum's inv. no: 2011/114 (A) - (KP 102)

Description: Square in form. Both faces are blank or illegible; flat profile. Corroded.

115) Islamic (10th-13th centuries)

Southern Anatolia

Dirham

Bronze, 11 x 11 x 3 mm, 2.85 g. Collected in Mersin and its vicinity and donated in 2010.

Museum's inv. no: 2011/115 (A) - (KP 101)

Description: Square in form. On one face, bird's eye; on the other face a circular stamp including kufic script (may be "Abd 7"); flat profile. Patinated and corroded. For similar script on one dirham weight see Kürkman 2003, no. 85 read as "Muhammed" and above, cat. nos. 104 and 107.

116) Islamic (10th-13th centuries)
Southern Anatolia
Dirham
Bronze, 9 x 10 x 4 mm, 2.85 g. Collected in Mersin and its vicinity and donated in 2010.
Museum's inv. no: 2011/116 (A) - (KP 107)
Description: Square in form. Bird's eye on the top and bottom; carinated profile. Patinated and corroded.
Cf. Hendin 2007, no. 449.

117) Islamic (10th-13th centuries)
Southern Anatolia
Dirham
Bronze, 10 x 11 x 3 mm, 2.80 g. Collected in Mersin and its vicinity and donated in 2010.
Museum's inv. no: 2011/117 (A) - (KP 100)
Description: Square in form. Large bird's eye on the top and bottom; flat profile. Patinated and corroded.
Cf. Hendin 2007, no. 449.

118) Islamic (10th-13th centuries)
Southern Anatolia
Dirham
Bronze, 10 x 10 x 3 mm, 2.80 g. Collected in Mersin and its vicinity and donated in 2010.
Museum's inv. no: 2011/118 (A) - (KP 104)
Description: Square in form. Bird's eye on the top and bottom; flat profile. Patinated and corroded.
Cf. Hendin 2007, no. 449.

119) Islamic (10th-13th centuries)
Southern Anatolia
Dirham
Bronze, 9 x 9 x 4 mm, 2.80 g. Collected in Mersin and its vicinity and donated in 2010.
Museum's inv. no: 2011/119 (A) - (KP 112)
Description: Hardly visible bird's eye on the top and bottom; carinated profile. Patinated and corroded.
Cf. Hendin 2007, no. 449.

120) Islamic (10th-13th centuries)
Southern Anatolia
Dirham
Bronze, 9 x 9 x 5 mm, 2.80 g. Collected in Mersin and its vicinity and donated in 2010.
Museum's inv. no: 2011/120 (A) - (KP 111)
Description: Square in form. Two bird's eye on one face and one bird's eye on the other; flat profile. Corroded.

Cat. no. 95

Cat. no. 107

Cat. no. 124

Cat. no. 127

Cat. no. 136

Cat. no. 159

Cat. no. 160

Cat. no. 161

121) Islamic (10th-13th centuries)

Southern Anatolia

Dirham

Bronze, 11 x 12 x 12 mm, 2.70 g. Collected in Mersin and its vicinity and donated in 2010.

Museum's inv. no: 2011/121 (A) - (KP 98)

Description: Square in form. Bird's eye on the top and bottom; carinated profile. Patinated and corroded.

Cf. Hendin 2007, no. 449.

122) Islamic (10th-13th centuries)

Southern Anatolia

Dirham

Bronze, 10 x 10 x 3 mm, 2.70 g. Collected in Mersin and its vicinity and donated in 2010.

Museum's inv. no: 2011/122 (A) - (KP 106)

Description: Square in form. Bird's eye on the top and bottom; flat profile. Patinated and corroded.

Cf. Hendin 2007, no. 449.

123) Islamic (10th-13th centuries)

Southern Anatolia

Half-dirham

Bronze, 9 x 9 x 3 mm, 1.30 g. Collected in Mersin and its vicinity and donated in 2010.

Museum's inv. no: 2011/123 (A) - (KP 115)

Description: Square in form. Illegible traces of Arabic letters on both faces.

124) Islamic (10th-13th centuries)

Southern Anatolia

Twenty-dirham

Bronze, 25 x 17 mm, 59.60 g. Collected in Mersin and its vicinity and donated in 2010.

Museum's inv. no: 2011/124 (A) - (KP 148)

Description: The weight has the form of two truncated cones with a distinct linear edge around the waist where the bases of the two truncated cones meet (barrel-like); very small centring point and two concentring circles (grooved) on the top and bottom; raised rim on the top only; profile blank but there are vertical linear lines; patinated and corroded.

Cf. Holland 1986, no. 70-71; Hendin 2007, no. 425; Holland 2009, nos. 113-116.

125) Islamic (10th-13th centuries)

Southern Anatolia

Ten-dirham

Bronze; 18 x 15 mm, 29.00 g. Collected in Mersin and its vicinity and donated in 2010.

Museum's inv. no: 2011/125 (A) - (KP 139)

Description: The weight has the form of two truncated cones with a distinct linear edge around the waist where the bases of the two truncated cones meet; concave bird's eye pattern on the top and bottom; raised rim on the top only; profile blank but there are vertical linear lines and faces; patinated and corroded.

Cf. Holland 1986, no. 5 (twenty-dirham, 58.50 g); Isler 1998, no. 34 (ten-dirham, 28.54 g); Holland 2009, no. 148.

126) Islamic (10th-13th centuries)

Southern Anatolia

Five-dirham

Bronze; 16 x 13 mm, 14.45 g. Collected in Mersin and its vicinity and donated in 2010.

Museum's inv. no: 2011/126 (A) - (KP 158)

Description: The weight has the form of two truncated cones with a distinct linear edge around the waist (*i.e.* carinated) where the bases of the two truncated cones meet; concave bird's eye pattern on the top and bottom; profile blank but there are vertical linear lines and faces. Patinated.

Cf. Holland 1986, no. 5 (twenty-dirham, 58.50 g); Isler 1998, no. 34 (ten-dirham, 28.54 g).

127) Islamic (10th-13th centuries)

Southern Anatolia

Twenty-dirham

Bronze; 25 x 19 mm, 59.00 g. Collected in Mersin and its vicinity and donated in 2010.

Museum's inv. no: 2011/127 (A) - (KP 127)

Description: The weight is a flattened sphere doubly truncated in form; punched on the top and bottom with a large bird's eye pattern; the main body of the weight is covered all over with close-set "bird's eyes". Patinated.

Cf. Qedar 1978, no. 113; Qedar 1979, no. 4186 ff.; Qedar 1983, no. 5172 (all misattributed to Byzantine); Holland 1986, no. 1; Hendin 2007, 412; Holland 2009, no. 123; Khamis 2010, no. 1 (early islamic).

128) Islamic (10th-13th centuries)

Southern Anatolia

Twenty-dirham

Bronze, 23 x 22 mm, 56.40 g. Collected in Mersin and its vicinity and donated in 2010.

Museum's inv. no: 2011/128 (A) - (KP 152)

Description: The weight is a flattened sphere doubly truncated in form; punched on the top and bottom with a bird's eye pattern (concentric circles); on the main body of the weight are three rows of bird's eyes but hardly visible. Patinated and corroded.

129) Islamic (AD 10th-13th centuries)

Southern Anatolia

Ten-dirham

Bronze; 20 x 15 mm; 29.55 g. Collected in Mersin and its vicinity and donated in 2010.

Museum's inv. no: 2011/129 (A) - (KP 142)

Description: The weight is a flattened sphere doubly truncated in form; punched on the top and bottom with a large and concave bird's eye pattern (concentric circles); on the main body of the weight are three rows of bird's eyes. Patinated and corroded.

This weight and the following weights are also similar in form (not aways in unit) with Qedar 1981, no.s. 137-140; Qedar 1983, nos. 5174, 5176 (both misattributed to Byzantine); Holland 1986, nos. 18 and Kürkman 2003, 98 and 101; Holland 2009, no. 160 ff.

130) Islamic (10th-13th centuries)

Southern Anatolia

Ten-dirham

Bronze, 19 x 15 mm, 29.45 g. Collected in Mersin and its vicinity and donated in 2010.

Museum's inv. no: 2011/130 (A) - (KP 130)

Description: The weight is a flattened sphere doubly truncated in form; punched on the top and bottom with a large and concave bird's eye pattern (concentric circles); on the main body of the weight are three rows of bird's eyes. Patinated and corroded.

131) Islamic (10th-13th centuries)

Southern Anatolia

Ten-dirham

Bronze, 20 x 15 mm, 29.35 g. Collected in Mersin and its vicinity and donated in 2010.

Museum's inv. no: 2011/131 (A) - (KP 137)

Description: The weight is a flattened sphere doubly truncated in form; punched on the top and bottom with a large and concave bird's eye pattern; on the main body of the weight are three rows of bird's eyes. Patinated and corroded.

132) Islamic (10th-13th centuries)

Southern Anatolia

Ten-dirham

Bronze, 19 x16 mm, 29.40 g. Collected in Mersin and its vicinity and donated in 2010.

Museum's inv. no: 2011/132 (A) - (KP 129)

Description: The weight is a flattened sphere doubly truncated in form; punched on the top and bottom with a large and concave bird's eye pattern; on the main body of the weight are three rows of bird's eyes. Patinated and corroded.

Cf. Kürkman 2003, 98 and 101.

133) Islamic (10th-13th centuries)

Southern Anatolia

Ten-dirham

Bronze, 20 x 16 mm, 29.30 g. Collected in Mersin and its vicinity and donated in 2010.

Museum's inv. no: 2011/133 (A) - (KP 138)

Description: The weight is a flattened sphere doubly truncated in form; punched on the top and bottom with a large and concave bird's eye pattern; the main body of the weight is covered all over with close-set bird's eyes. Patinated and corroded.

134) Islamic (10th-13th centuries)

Southern Anatolia

Ten-dirham

Bronze, 20 x15 mm, 29.10 g. Collected in Mersin and its vicinity and donated in 2010.

Museum's inv. no: 2011/134 (A) - (KP 141)

Description: The weight is a flattened sphere doubly truncated in form; punched central dot on the top; bottom is flat; the main body of the weight is covered all over with bird's eyes. Patinated and corroded.

135) Islamic (10th-13th centuries)

Southern Anatolia

Ten-dirham

Bronze, 20 x 15 mm, 29.15 g. Collected in Mersin and its vicinity and donated in 2010.

Museum's inv. no: 2011/135 (A) - (KP 136)

Description: The weight is a flattened sphere doubly truncated in form; punched on the top and bottom with a large bird's eye pattern; on the main body of the weight are three rows of bird's eyes. Patinated.

Cf. Kürkman 2003, 98 and 101.

136) Islamic (10th-13th centuries)

Southern Anatolia

Ten-dirham

Bronze, 20 x 15 mm, 28.20 g. Collected in Mersin and its vicinity and donated in 2010.

Museum's inv. no: 2011/136 (A) - (KP 132)

Description: The weight is a flattened sphere doubly truncated in form; punched on the top and bottom with a large bird's eye pattern; on the main body of the weight are three rows of bird's eyes. Patinated.

Cf. Kürkman 2003, 98 and 101.

137) Islamic (10th-13th centuries)

Southern Anatolia

Ten-dirham

Bronze, 20 x 16 mm, 28.85 g. Collected in Mersin and its vicinity and donated in 2010.

Museum's inv. no: 2011/137 (A) - (KP 131)

Description: The weight is a flattened sphere doubly truncated in form; punched on the top and bottom with a concave bird's eye pattern; the main body of the weight is covered all over with close-set bird's eyes. Patinated.

138) Islamic (10th-13th centuries)

Southern Anatolia

Ten-dirham

Bronze, 19 x 15 mm, 28.70 g. Collected in Mersin and its vicinity and donated in 2010.

Museum's inv. no: 2011/138 (A) - (KP 140)

Description: The weight is a flattened sphere doubly truncated in form; punched on the top and bottom with a large bird's eye pattern (concentric circles); on the main body of the weight are three rows of bird's eyes. Patinated and corroded.

Cf. Kürkman 2003, 98 and 101.

139) Islamic (10th-13th centuries)

Southern Anatolia

Ten-dirham

Bronze, 19 x15 mm, 27.75 g. Collected in Mersin and its vicinity and donated in 2010.

Museum's inv. no: 2011/139 (A) - (KP 135)

Description: The weight is a flattened sphere doubly truncated in form; punched on the top and bottom with a large and concave bird's eye pattern; on the main body of the weight are three rows of bird's eyes.

Cf. Kürkman 2003, 98 and 101.

140) Islamic (10th-13th centuries)

Southern Anatolia

Ten-dirham

Bronze, 20 x15 mm, 27.65 g. Collected in Mersin and its vicinity and donated in 2010.

Museum's inv. no: 2011/140 (A) - (KP 134)

Description: The weight is a flattened sphere doubly truncated in form; punched on the top and bottom with a large bird's eye pattern; on the main body of the weight are three rows of bird's eyes. Heavily corroded.

Cf. Kürkman 2003, 98 and 101.

141) Islamic (10th-13th centuries)

Southern Anatolia

Ten-dirham

Bronze, 20 x 15 mm, 27.65 g. Collected in Mersin and its vicinity and donated in 2010.

Museum's inv. no: 2011/141 (A) - (KP 128)

Description: The weight is a flattened sphere in form; punched on the top and bottom with a concave bird's eye pattern; on the main body of the weight are three rows of bird's eyes. Patinated and corroded.

Cf. Kürkman 2003, 98 and 101.

142) Islamic (10th-13th centuries)

Southern Anatolia

Ten-dirham (light)

Bronze, 18 x 15 mm, 26.40 g. Collected in Mersin and its vicinity and donated in 2010.

Museum's inv. no: 2011/142 (A) - (KP 133)

Description: The weight is a flattened sphere in form; punched on the top and bottom with a large and concave bird's eye pattern; on the main body of the weight are three rows of bird's eyes. Patinated and corroded.

Cf. Kürkman 2003, 98 and 101.

143) Islamic (10th-13th centuries)

Southern Anatolia

Five-dirham

Bronze, 17 x 14 mm, 15.20 g. Collected in Mersin and its vicinity and donated in 2010.

Museum's inv. no: 2011/143 (A) - (KP 154)

Description: The weight is a flattened sphere in form; punched on the top and bottom with a large and concave bird's eye pattern; on the main body of the weight are three rows of bird's eyes but hardly visible. Heavily corroded.

Cf. Hendin 2007, 417.

144) Islamic (10th-13th centuries)

Southern Anatolia

Five-dirham

Bronze, 15 x 13 mm, 14.80 g. Collected in Mersin and its vicinity and donated in 2010.

Museum's inv. no: 2011/144 (A) - (KP 157)

Description: The weight is a flattened sphere doubly truncated in form; punched on the top and bottom with a large and concave bird's eye pattern; on the main body of the weight are three rows of bird's eyes. Patinated.

Cf. Hendin 2007, 417.

145) Islamic (10th-13th centuries)

Southern Anatolia)

Five-dirham

Bronze, 15 x13 mm, 14.80 g. Collected in Mersin and its vicinity and donated in 2010.

Museum's inv. no: 2011/145 (A) - (KP 167)

Description: The weight is a flattened sphere in form; punched on the top and bottom with a large and concave bird's eye pattern; on the main body of the weight are three rows of bird's eyes. Patinated and corroded.

146) Islamic (10th-13th centuries)

Southern Anatolia

Five-dirham

Bronze, 15 x 13 mm, 14.75 g. Collected in Mersin and its vicinity and donated in 2010.

Museum's inv. no: 2011/146 (A) - (KP 155)

Description: The weight is a flattened sphere doubly truncated in form; punched on the top and bottom with a large and concave bird's eye pattern; on the main body of the weight are three rows of bird's eyes.

Cf. Kürkman 2003, 93; Hendin 2007, 418.

147) Islamic (10th-13th centuries)

Southern Anatolia

Five-dirham

Bronze, 15 x 13 mm, 14.70 g. Collected in Mersin and its vicinity and donated in 2010.

Museum's inv. no: 2011/147 (A) - (KP 166)

Description: The weight is a flattened sphere doubly truncated in form; punched on the top and bottom with a large and concave bird's eye pattern; on the main body of the weight are three rows of bird's eyes. Patinated and corroded.

Cf. Kürkman 2003, 93; Hendin 2007, 417.

148) Islamic (10th-13th centuries)

Southern Anatolia

Five-dirham

Bronze, 15 x 13 mm, 14.65 g. Collected in Mersin and its vicinity and donated in 2010.

Museum's inv. no: 2011/148 (A) - (KP 159)

Description: The weight is a flattened sphere doubly truncated in form; punched on the top and bottom with a large and concave bird's eye pattern; on the main body of the weight are three rows of bird's eyes. Patinated.

Cf. Kürkman 2003, 93; Hendin 2007, 417.

149) Islamic (10th-13th centuries)

Southern Anatolia

Five-dirham

Bronze, 16 x 12 mm, 14.55 g. Collected in Mersin and its vicinity and donated in 2010.

Museum's inv. no: 2011/149 (A) - (KP 153)

Description: The weight is a flattened sphere doubly truncated in form; punched on the top and bottom with a large bird's eye pattern; on the main body of the weight are three rows of bird's eyes. Patinated.

150) Islamic (10th-13th centuries)

Southern Anatolia

Five-dirham

Bronze, 15 x 12 mm, 14.50 g. Collected in Mersin and its vicinity and donated in 2010.

Museum's inv. no: 2011/150 (A) - (KP 163)

Description: The weight is a flattened sphere doubly truncated in form; a distinct linear edge (carinate) around the waist; punched centring point and circular groove near the edge on the top and bottom; on the main body of the weight are two rows of bird's eyes. Patinated.

151) Islamic (10th-13th centuries)

Southern Anatolia

Five-dirham

Bronze, 15 x 13 mm, 14.50 g. Collected in Mersin and its vicinity and donated in 2010.

Museum's inv. no: 2011/151 (A) - (KP 156)

Description: The weight is a flattened sphere in form; top and bottom blank; on the body of the weight are three rows of bird's eyes. Patinated and corroded.

152) Islamic (10th-13th centuries)

Southern Anatolia

Five-dirham

Bronze, 15 x 12 mm, 14.50 g. Collected in Mersin and its vicinity and donated in 2010.

Museum's inv. no: 2011/152 (A) - (KP 168)

Description: The weight is a flattened sphere; a distinct linear edge (carinate) around the waist; the polar facets (top and bottom) are large and blank; on the main body of the weight are two rows of bird's eyes. Patinated and corroded.

153) Islamic (10th-13th centuries)

Southern Anatolia

Five-dirham

Bronze, 16 x 12 mm,14.45 g. Collected in Mersin and its vicinity and donated in 2010.

Museum's inv. no: 2011/153 (A) - (KP 161)

Description: The weight is a flattened sphere doubly truncated in form; punched on the top and bottom with a small bird's eye pattern; on the main body of the weight are three rows of bird's eyes. Patinated and corroded.

Cf. Hendin 2007, 417.

154) Islamic (10th-13th centuries)

Southern Anatolia

Five-dirham

Bronze, 15 x 13 mm, 14.35 g. Collected in Mersin and its vicinity and donated in 2010.

Museum's inv. no: 2011/154 (A) - (KP 162)

Description: The weight is a flattened sphere doubly truncated in form; punched on the top and bottom with a bird's eye pattern; on the main body of the weight are three rows of bird's eyes. Patinated.

Cf. Kürkman 2003, no. 93; Hendin 2007, 418.

155) Islamic (10th-13th centuries)

Southern Anatolia

Five-dirham

Bronze, 15 x 12 mm, 14.35 g. Collected in Mersin and its vicinity and donated in 2010.

Museum's inv. no: 2011/155 (A) - (KP 164)

Description: The weight is a flattened sphere doubly truncated in form; punched on the top and bottom with a large bird's eye pattern; on the main body of the weight are three rows of bird's eyes. Patinated and corroded.

Cf. Kürkman 2003, no. 93; Hendin 2007, 418.

156) Islamic (10th-13th centuries)

Southern Anatolia

Five-dirham

Bronze, 15 x12 mm, 14.05 g. Collected in Mersin and its vicinity and donated in 2010.

Museum's inv. no: 2011/156 (A) - (KP 165)

Description: The weight is a flattened sphere doubly truncated in form; punched on the top and bottom with three irregular small bird's eye pattern; on the main body of the weight are three rows of bird's eyes. Patinated and corroded.

157) Islamic (10th-13th centuries)

Southern Anatolia

Five-dirham

Bronze, 15 x 13 mm, 13.90 g. Collected in Mersin and its vicinity and donated in 2010.

Museum's inv. no: 2011/157 (A) - (KP 160)

Description: The weight is a flattened sphere doubly truncated in form; punched on the top and bottom with a small bird's eye pattern; on the main body of the weight are three rows of bird's eyes. Patinated and corroded.

Cf. Kürkman 2003, no. 93; Hendin 2007, no. 418.

158) Islamic? (10th-14th centuries?)

Southern Anatolia

Ten-dirham?

Bronze, 30 x 6 mm, 24.30 g. Collected in Mersin and its vicinity and donated in 2010.

Museum's inv. no: 2011/158 (A) - (KP 146)

Description: Hexagonal in form; smooth and well proportioned; top and bottom are blank or illegible. The weight of this piece corresponds also approximately to Byzantine six-nomismata or one uncia. Corroded.

Cf. Isler 1998, nos. 20-21; Nenci 2001, no. 4.

159) Islamic, Ottoman (18th-19th centuries)

Southern Anatolia

Twenty-dirham

Bronze, 27 x 14 mm, 62.40 g. Collected in Mersin and its vicinity and donated in 2010.

Museum's inv. no: 2011/159 (A) - (KP 147)

Description: Octogonal in form. On the top, illegible arabic letters or motif within stamp (tughra?); bottom, illegible. Patinated.

For similar form cf. Balog 1970, nos. 34-35; Custurea 1998, no. 1; Kürkman 2003, nos. 200, 207-208, 221-225, 231-232, 326, 328, 448.

160) Islamic, Ottoman (18th-19th centuries)

Southern Anatolia

Ten-dirham

Brass, 20 x 13 mm, 31.55 g. Collected in Mersin and its vicinity and donated in 2010.

Museum's inv. no: 2011/160 (A) - (KP 150)

Description: Octogonal in form; top and bottom are blank. Its mass corresponds to ten-dirham.

161) Islamic, Ottoman (18th-19th centuries or may be earlier)

Southern Anatolia

Twenty-dirham

Bronze, 25 mm, 61.50 g. Collected in Mersin and its vicinity and donated in 2010.

Museum's inv. no: 2011/161 (A) - (KP 151)

Description: Polyhedron; faces are blank. Patinated.

Museum's inv no. KP 151

Cf. Kisch 1965, no. 53; Custurea 1995-1996, no. 6 ff; Custurea 1998, no. 2; Custurea 2002-2003, no. 7 ff.

LIST OF CONCORDANCE

Inv. no.	Cat. no.	Inv. no.	Cat. no.	Inv. no.	Cat. no.	Inv. no.	Cat. no.	Inv. no.	Cat. no.
2011/1 (A)	1	2011/34 (A)	34	2011/66 (A)	66	2011/98 (A)	98	2011/130 (A)	130
2011/2 (A)	2	2011/35 (A)	35	2011/67 (A)	67	2011/99 (A)	99	2011/131 (A)	131
2011/3 (A)	3	2011/36 (A)	36	2011/68 (A)	68	2011/100 (A)	100	2011/132 (A)	132
2011/4 (A)	4	2011/37 (A)	37	2011/69 (A)	69	2011/101 (A)	101	2011/133 (A)	133
2011/5 (A)	5	2011/38 (A)	38	2011/70 (A)	70	2011/102 (A)	102	2011/134 (A)	134
2011/6 (A)	6	2011/39 (A)	39	2011/71 (A)	71	2011/103 (A)	103	2011/135 (A)	135
2011/7 (A)	7	2011/40 (A)	40	2011/72 (A)	72	2011/104 (A)	104	2011/136 (A)	136
2011/8 (A)	8	2011/41 (A)	41	2011/73 (A)	73	2011/105 (A)	105	2011/137 (A)	137
2011/9 (A)	9	2011/42 (A)	42	2011/74 (A)	74	2011/106 (A)	106	2011/138 (A)	138
2011/10 (A)	10	2011/43 (A)	43	2011/75 (A)	75	2011/107 (A)	107	2011/139 (A)	139
2011/11 (A)	11	2011/44 (A)	44	2011/76 (A)	76	2011/108 (A)	108	2011/140 (A)	140
2011/12 (A)	12	2011/45 (A)	45	2011/77 (A)	77	2011/109 (A)	109	2011/141 (A)	141
2011/13 (A)	13	2011/46 (A)	46	2011/78 (A)	78	2011/110 (A)	110	2011/142 (A)	142
2011/14 (A)	14	2011/47 (A)	47	2011/79 (A)	79	2011/111 (A)	111	2011/143 (A)	143
2011/15 (A)	15	2011/48 (A)	48	2011/80 (A)	80	2011/112 (A)	112	2011/144 (A)	144
2011/16 (A)	16	2011/49 (A)	49	2011/81 (A)	81	2011/113 (A)	113	2011/145 (A)	145
2011/17 (A)	17	2011/50 (A)	50	2011/82 (A)	82	2011/114 (A)	114	2011/146 (A)	146
2011/18 (A)	18	2011/51 (A)	51	2011/83 (A)	83	2011/115 (A)	115	2011/147 (A)	147
2011/19 (A)	19	2011/52 (A)	52	2011/84 (A)	84	2011/116 (A)	116	2011/148 (A)	148
2011/20 (A)	20	2011/53 (A)	53	2011/85 (A)	85	2011/117 (A)	117	2011/149 (A)	149
2011/21 (A)	21	2011/54 (A)	54	2011/86 (A)	86	2011/118 (A)	118	2011/150 (A)	150
2011/22 (A)	22	2011/55 (A)	55	2011/87 (A)	87	2011/119 (A)	119	2011/151 (A)	151
2011/23 (A)	23	2011/56 (A)	56	2011/88 (A)	88	2011/120 (A)	120	2011/152 (A)	152
2011/24 (A)	24	2011/57 (A)	57	2011/89 (A)	89	2011/121 (A)	121	2011/153 (A)	153
2011/25 (A)	25	2011/58 (A)	58	2011/90 (A)	90	2011/122 (A)	122	2011/154 (A)	154
2011/26 (A)	26	2011/59 (A)	59	2011/91 (A)	91	2011/123 (A)	123	2011/155 (A)	155
2011/27 (A)	27	2011/60 (A)	60	2011/92 (A)	92	2011/124 (A)	124	2011/156 (A)	156
2011/28 (A)	28	2011/61 (A)	61	2011/93 (A)	93	2011/125 (A)	125	2011/157 (A)	157
2011/29 (A)	29	2011/62 (A)	62	2011/94 (A)	94	2011/126 (A)	126	2011/158 (A)	158
2011/30 (A)	30	2011/63 (A)	63	2011/95 (A)	95	2011/127 (A)	127	2011/159 (A)	159
2011/31 (A)	31	2011/64 (A)	64	2011/96 (A)	96	2011/128 (A)	128	2011/160 (A)	160
2011/32 (A)	32	2011/65 (A)	65	2011/97 (A)	97	2011/129 (A)	129	2011/161 (A)	161
2011/33 (A)	33								